AF575909

DISCOVERING STEM

SCREWS

Coral Martincavage

EZ READERS

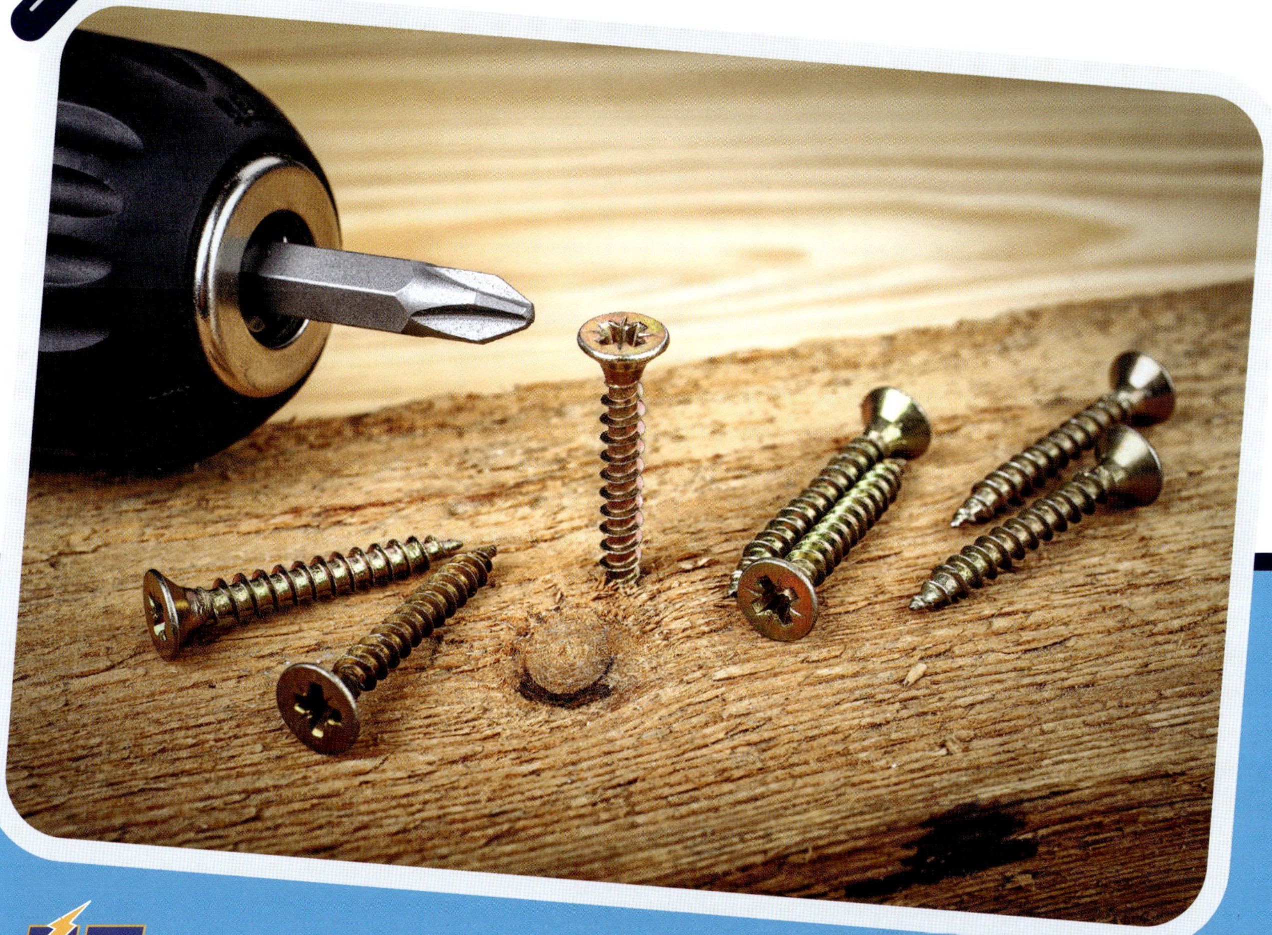

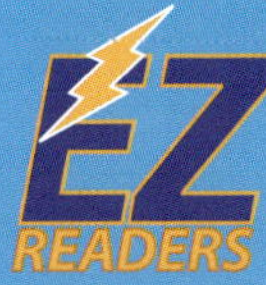

Creating Young Nonfiction Readers

EZ Readers lets children delve into nonfiction at beginning reading levels. Young readers are introduced to new concepts, facts, ideas, and vocabulary.

Tips for Reading Nonfiction with Beginning Readers

Talk about Nonfiction
Begin by explaining that nonfiction books give us information that is true. The book will be organized around a specific topic or idea, and we may learn new facts through reading.

Look at the Parts
Most nonfiction books have helpful features. Our *EZ Readers* include a Contents page, an Index, and color photographs. Share the purpose of these features with your reader.

Contents
Located at the front of a book, the Contents displays a list of the big ideas within the book and where to find them.

Index
An Index is an alphabetical list of topics and the page numbers where they are found.

Photos/Charts
A lot of information can be found by "reading" the charts and photos found within nonfiction text. Help your reader learn more about the different ways information can be displayed.

With a little help and guidance about reading nonfiction, you can feel good about introducing a young reader to the world of *EZ Readers* nonfiction books.

Mitchell Lane
PUBLISHERS

mitchelllane.com

2001 SW 31st Avenue
Hallandale, FL 33009
www.mitchelllane.com

First Edition, 2021.

Author: Coral Martincavage
Designer: Ed Morgan
Editor: Morgan Brody

Names/credits:
Title: Screws / by Coral Martincavage
Description: Hallandale, FL :
Mitchell Lane Publishers, [2021]

Series: Discovering STEM
Library bound ISBN: 978-1-68020-585-5
eBook ISBN: 978-1-68020-586-2

EZ readers is an imprint of Mitchell Lane Publishers.

Photo credits: Freepik.com, Shutterstock, pp. 8--9 Photo by Chris Yates on Unsplash, pp. 16-17 Photo by Caleb Woods on Unsplash,

CONTENTS

Words in **bold** can be found in the Glossary.

Let’s build a house.
What do we need?
We need wood and **screws**.

2m
M14

What is a screw? A screw is a small metal pin. It has **threads** all around. The tip is a sharp point that digs into the wood.

Some threads are close together. Others are far apart.

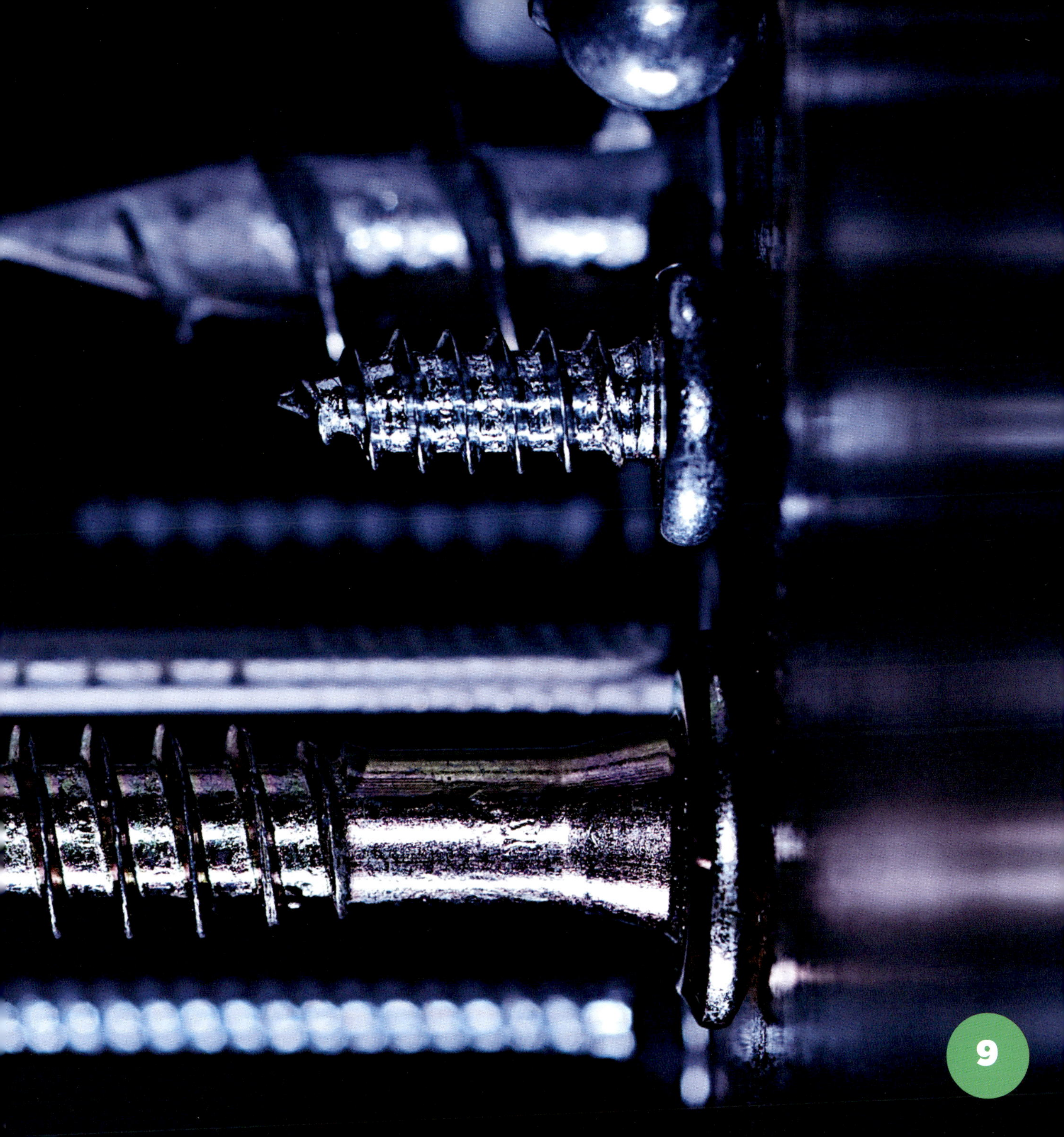

10

The threads help the screw go in or out. A **screwdriver** or wrench can help. Make it tight!

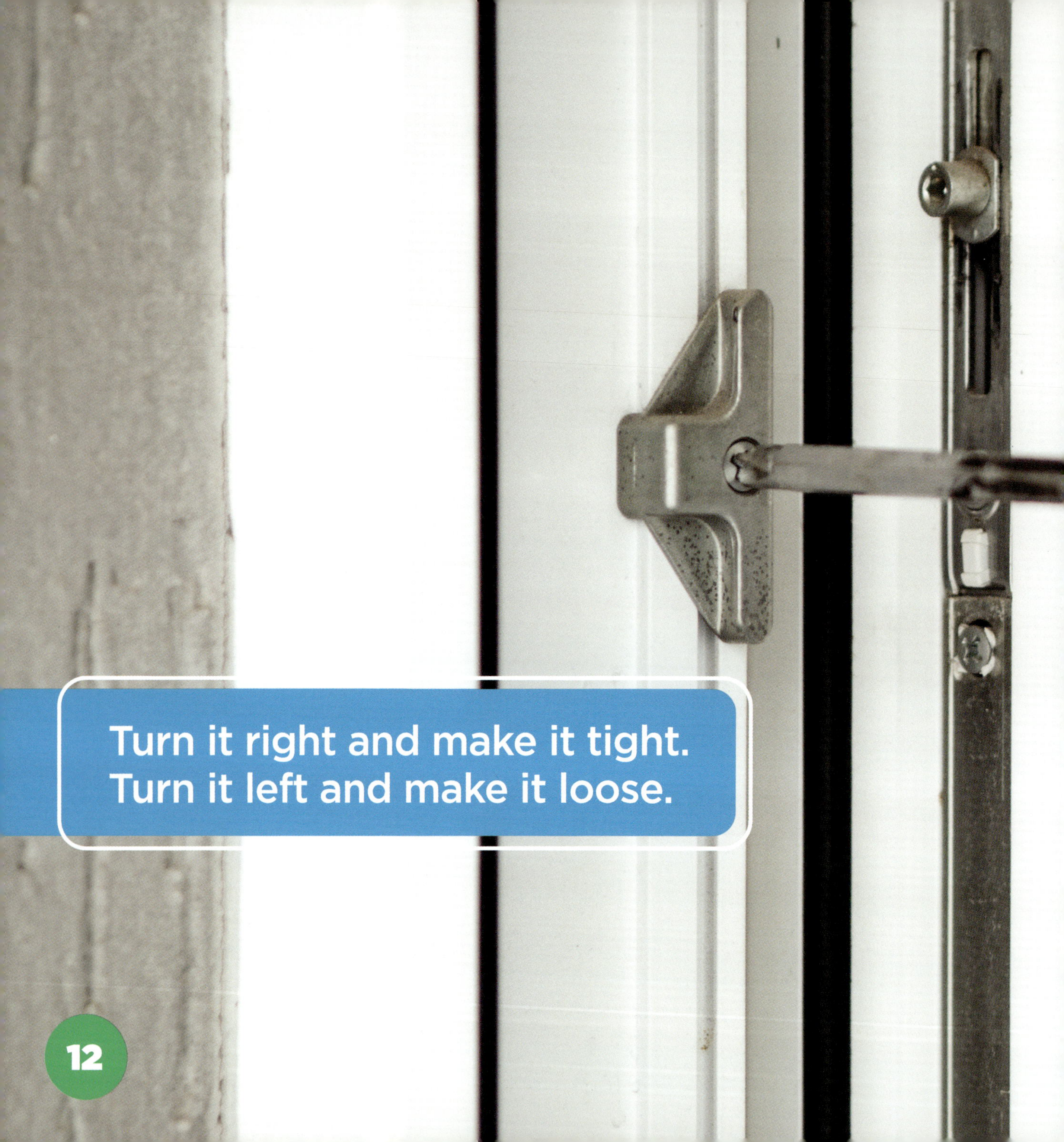

Turn it right and make it tight.
Turn it left and make it loose.

More threads need more turns, or **rotations**. Less threads need less turns, but they use more **force**. A force is a push or a pull.

It all takes **effort**. Effort is energy needed to do something.

18

A screw is a **simple machine**.
It holds things together.
Twist and push!

Screws are small but make a big difference. They help us in many ways!

INTERESTING FACTS

Did you know that many of the carpenter tools we use today, such as: the chisel, the plane, saw, level and screws were invented before the Roman Age?

Early screws had to be handmade and no two screws were ever alike.

More than 200 billion screws are used every year across the United States.

Screws, originally, weren't even created to fasten things. The first use of screws was for pressing grapes and olives to get their oils.

There are many kinds of screws, including wood screws, machine screws, hex cap screws, lag screws, as well as other fasteners and bolts.

Screws are used with a screwdriver. The screw actually came before the screwdriver. The first screws were probably tightened with a knife. The flathead and Phillips are the most popular screwdrivers.

GLOSSARY

effort
Physical energy needed to do something

force
A push or pull in a specific direction

rotation
A turn

screw
A short, sharp-pointed metal pin used to join things together

screwdriver
A tool for turning screws

simple machine
A device that makes work easier because it uses less effort to move an object

threads
Spiraling grooves on a screw

SOURCES

Simple Machines to the Rescue by Sharon Thales (Capstone Press, 2007).

Screws by Lyn Sirota and Reginald Butler (The Child's World, 2014).

FURTHER READING

Drew the Screw by Mattia Cerato (Holiday House, 2016).

Screws to the Rescue by Sharon Thales (Capstone Press, 2007).

Tools Rule! By Aaron Meshon (Atheneum Books for Young Readers, 2014).

INDEX

ABOUT THE AUTHOR

One of **Coral Martincavage's** favorite snacks is chocolate chip cookies. Each time she twists and pulls the cookie jar, she is reminded of how important screws are. She hopes this book inspires young readers to find ways that screws are useful in everyday life.